T0097189

Published by Ice House Books

Written and Designed by Smart Design Studio

Photography Shutterstock.com – for individual credits see back page

Ice House Books is an imprint of Half Moon Bay Limited
The Ice House, 124 Walcot Street, Bath, BA1 5BG
www.icehousebooks.co.uk

ISBN 978-1-912867-66-0

Printed in China

IFLSCIENCE!
SEVERE WEATHER
PRECIOUS PLANET
ICE HOUSE BOOKS

PLASTIC-GOBBLING MICROBES

MICROSCOPIC MARINE BACTERIA IS ON OUR SIDE!

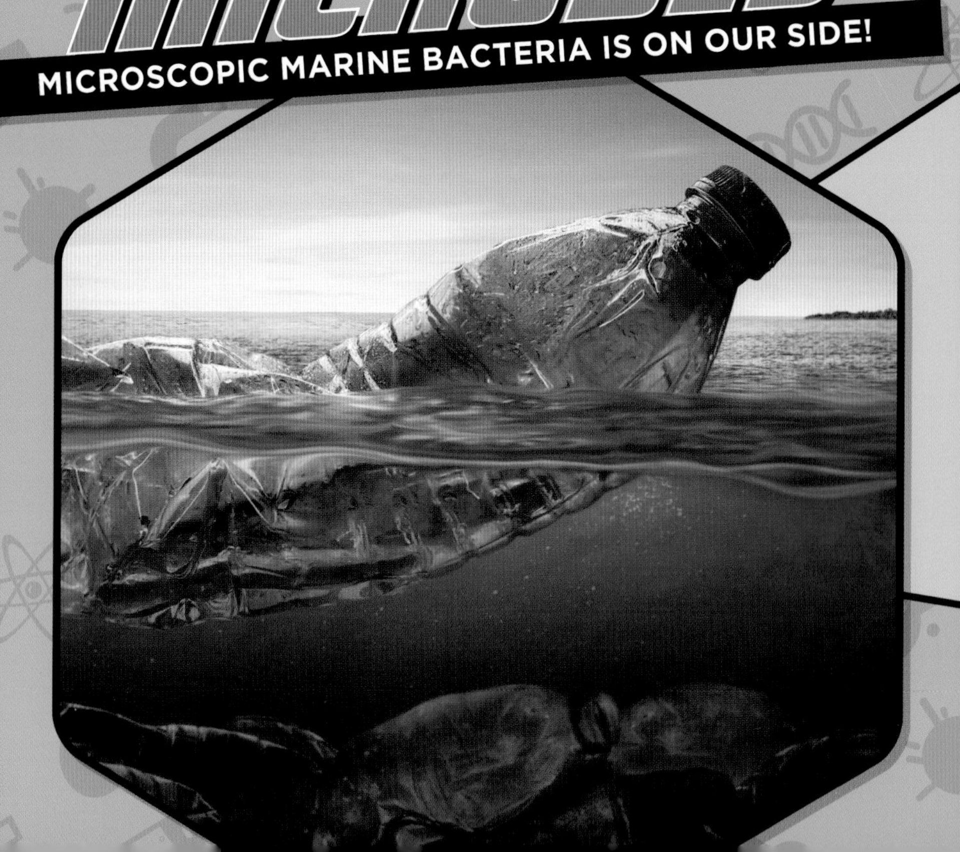

A team of scientists has recently revealed some mind-blowing findings. Evidence suggests that certain teeny-tiny microbes are evolving to break down plastic over many months.

As tasty as plastic is to some microorganisms, sadly they can't munch fast enough to keep up with the epic levels of plastic we produce and throw away.

MICROBES TO THE RESCUE!

#TRASHTAG CHALLENGE

SOCIAL MEDIA SAVES THE PLANET.

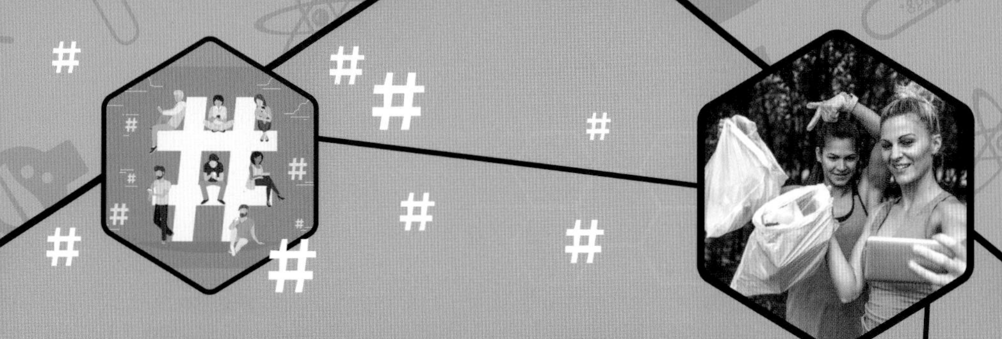

Every now and again an Internet craze appears that isn't just an excuse for people to look like idiots dancing next to a moving car *(We're here looking at you, #InMyFeelings)*.

We may be #soover lame social media challenges, but here is one planet-saving fad we can totally get behind.

From forest floors to sandy beaches, people are turning up in their thousands to take before and after pics of epic clean-up quests.

GET ONLINE AND SEARCH #TRASHTAG...

RADICAL RECYCLING

NORWAY IS SHOWING THE REST OF THE WORLD HOW IT'S DONE!

Norwegian citizens are charged a small 'mortgage' on plastic bottles, which they can claim back if they recycle their used bottles in special machines all over the country.

A MASSIVE 97% of plastic bottles are now recycled across the country!

ROOKS TO THE RESCUE!

BIRDS ARE PUTTING US TO SHAME!

A theme park in Western France has trained its resident rooks to swoop down and collect litter and cigarette butts from the ground.

There's no reason why this innovative approach to litter-picking couldn't work on a mass scale. Magpies, crows, ravens and rooks are all incredibly intelligent....

WINGED WASTE WARRIORS!

STUCK FOR IDEAS?

HERE ARE SOME EASY AND INNOVATIVE WAYS TO REDUCE, REUSE AND RECYCLE...

- **Poke some holes in a plastic bottle, fill it with seeds and string it up in the garden. Ta-da! You've made a bird feeder!**
- Use your old, ripped and stained t-shirts as dish rags and cleaning cloths. Cut down on paper towels!
- **Glass food jars make great storage containers for food, jewellery, stationery and any odds and ends you have lying around.**
- Start a herb garden with two empty 2 lt bottles.
- **Create a recycled plastic bottle jet pack for kids.**
- Tip a glass jar upside down and create a tiny terrarium.

JOIN THE GREEN SIDE...

MAN-MADE ISLANDS

...COULD SAVE THE PLANET.

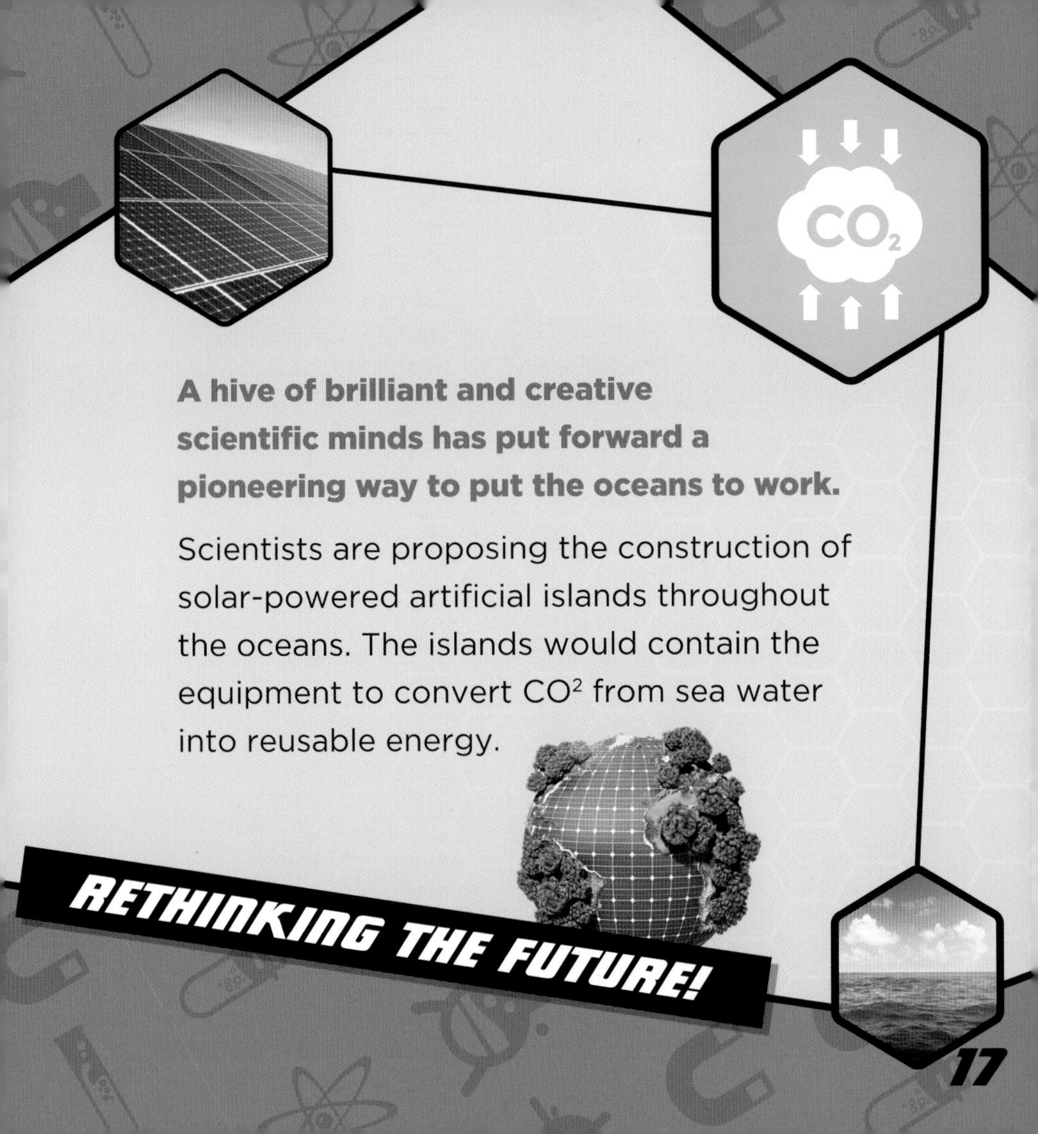

A hive of brilliant and creative scientific minds has put forward a pioneering way to put the oceans to work.

Scientists are proposing the construction of solar-powered artificial islands throughout the oceans. The islands would contain the equipment to convert CO^2 from sea water into reusable energy.

RETHINKING THE FUTURE!

EAT ORGANIC

PESTICIDES ARE GROSS AND PRODUCE GREENHOUSE GASES.

Fancy a salad with a dressing of snail pellets, weed killer and insect growth regulators?

NO?

Neither does the planet, so buy organic.

GOOD FOR NATURE, GOOD FOR YOU!

POWER TO THE PEOPLE!

IT'S NOT JUST THE SEA LEVELS THAT ARE RISING, BUT THE GENERAL PUBLIC TOO...

Politicians ignoring climate change, beware!

A group of residents from the Torres Strait Islands near Australia are making a case against their government to the United Nations. The residents say the catastrophic flood threat caused by the rising sea level is breaching their human rights.

MARIANA TRENCH

PLASTIC HAS BEEN FOUND IN THE DEEPEST PART OF THE OCEAN. ST'S JUST GOT REAL!**

But doing our bit to protect our oceans really is as simple as A, B, C...

A – Avoid products with microbeads

B – Ban the use of single-use plastic in your home

C – Clean up your litter!

RECYCLING ISN'T ROCKET SCIENCE!

GET OFF YOUR a**!

PUT DOWN THE CAR KEYS...

The average car produces all kinds of toxic gases, which are bad for the environment and our health. Carbon monoxide, hydrocarbons, nitrogen oxides ... the less we pump into the air the better.

So, try to drive smart – take fewer trips, carpool and just get off your backside when you can!

TODAY'S ACTIONS ARE TOMORROW'S RESULTS...

VIVE LA FRANCE!

CALLING ALL FUTURE SCIENTISTS...

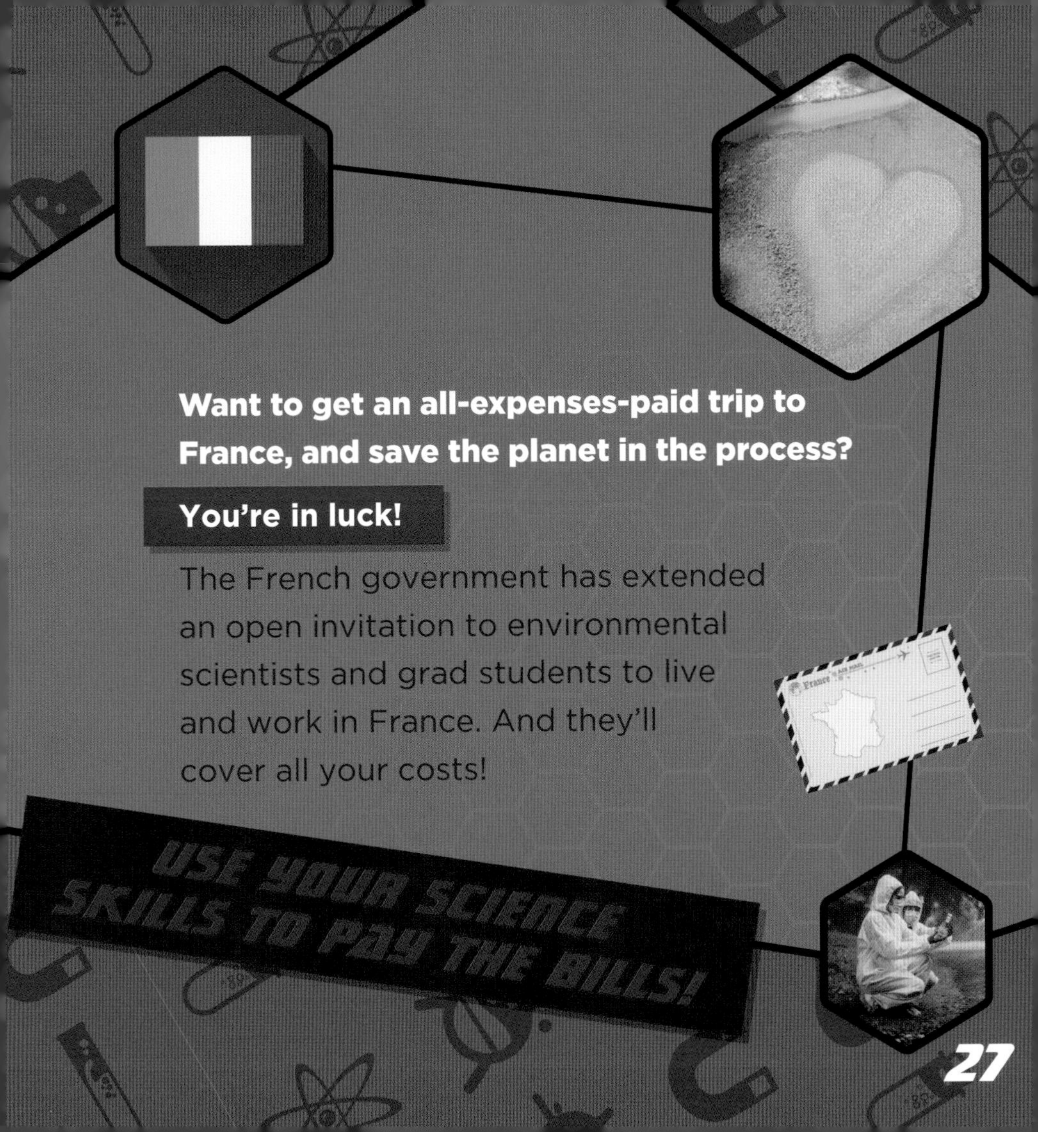

Want to get an all-expenses-paid trip to France, and save the planet in the process?

You're in luck!

The French government has extended an open invitation to environmental scientists and grad students to live and work in France. And they'll cover all your costs!

USE YOUR SCIENCE SKILLS TO PAY THE BILLS!

PLANET-PROTECTING P**S-UPS!

TOP FIVE TIPS FOR SUSTAINABLE DRINKING...

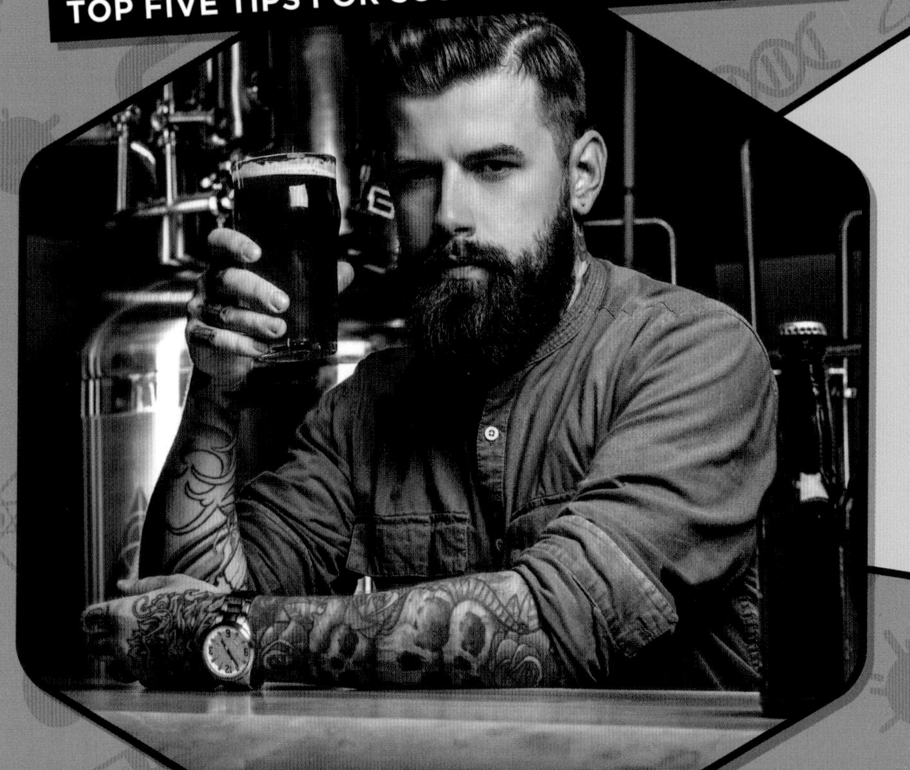

1. **Buy organic beer, wine and spirits.**
2. Mind your mixers – choose organic and locally-sourced.
3. **Buy local – the less distance booze has travelled, the smaller the carbon footprint.**

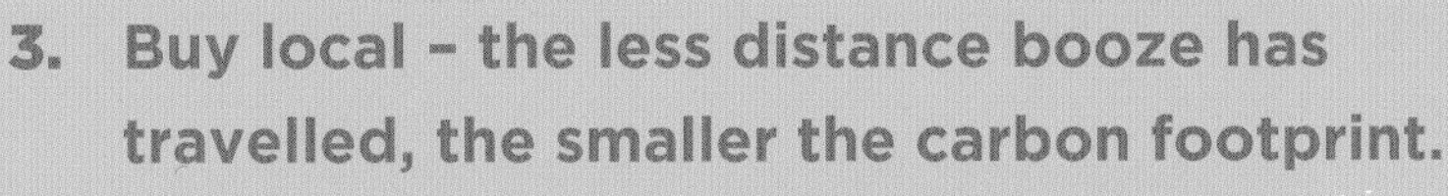

4. Sustainability – choose brewing and fermenting methods that are environmentally sustainable.
5. **Consider the containers – glass and aluminium can be recycled. Bottle caps cannot!**

SAVE THE HEDGEHOGS!

HEDGEHOGS ARE DYING OUT IN ALARMING NUMBERS...

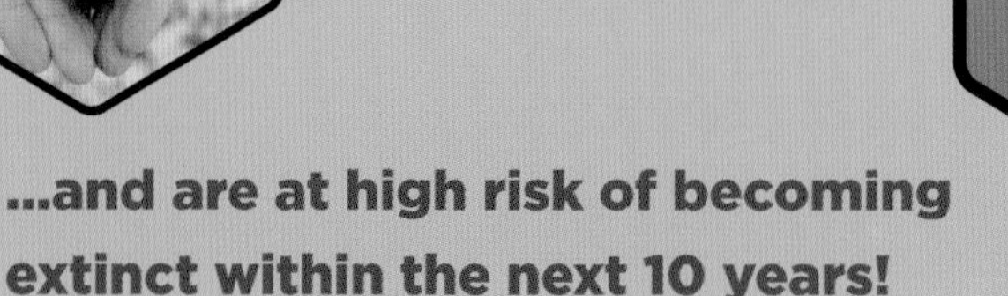

...and are at high risk of becoming extinct within the next 10 years!

Here are a few simple ways to help our favourite spikey friends...

- Don't use pesticides in your garden
- **Check before strimming**
- Pick up litter
- **Be careful with bonfires**
- Create a 'wild corner' in your garden

CONSERVATION STARTS AT HOME!

THE WORLD'S LARGEST WIND FARM

...IS IN THE UK!

That's right!

When it comes to renewable energy, we're not just blowing a lot of hot air.

The London array of 87 turbines is just off the coast of Kent and Essex and produces enough energy to fuel 590,000 homes!

FACT:
EACH TURBINE STANDS MORE THAN 600 FEET TALL, MORE THAN TWICE THE HEIGHT OF THE STATUE OF LIBERTY!

SCANDI POWER

DENMARK IS OFFICIALLY THE MOST CLIMATE-FRIENDLY COUNTRY IN THE WORLD!

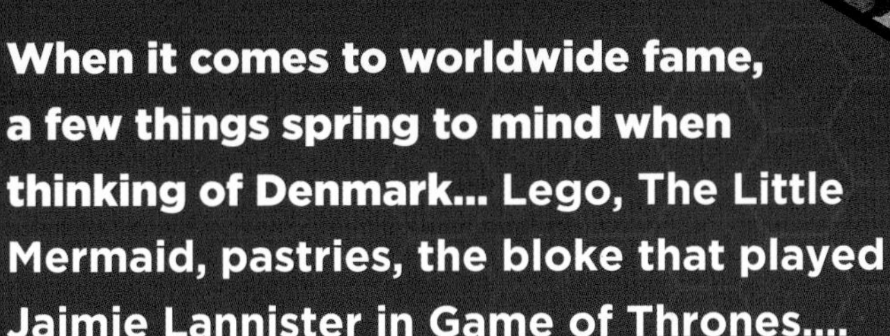

When it comes to worldwide fame, a few things spring to mind when thinking of Denmark... Lego, The Little Mermaid, pastries, the bloke that played Jaimie Lannister in Game of Thrones....

But now the Scandinavian country has a new claim to fame – it's officially the world's 'greenest' country.

DENMARK IS ON COURSE TO BE 100% FOSSIL FUEL INDEPENDENT BY 2050!

BEE GOOD

JUST A SPOONFUL OF SUGAR...

...CAN **HELP** REVIVE **A** DYING **BEE.**

And we like bees, bees are good. They pollinate plants and without them we'd be in trouble.

So next time you see a tired bee, dissolve two tablespoons of sugar in some water and leave it within the bee's reach.

You're no fool. You're not going to argue with David Attenborough....

GRETA THUNBERG

THE SCHOOL GIRL SAVING THE PLANET!

Teenage climate activist, Greta Thunberg, is inspiring a generation of teenagers to act on climate change.

The **#FridaysForFuture** demonstrations have seen mass walkouts by school children over the globe.

'WE CAN'T SAVE THE WORLD BY PLAYING BY THE RULES, BECAUSE THE RULES HAVE TO BE CHANGED.'
GRETA THUNBERG

CRUMBS IN A CAN

DON'T KNOW WHAT TO DO WITH YOUR OLD BREAD CRUSTS?

We know! Make them into beer!

The baking and brewing industries have sat side by side since the days of ancient Rome.

But an innovative British brewing company has resurrected the long-lost skill set of turning bread into beer!

BIN THE FOOD BIN!

HOW TO REDUCE FOOD WASTE IN FIVE EASY STEPS...

1. **Shop smart** – make a list of the items you need and stick to it.
2. **Get down with pickling** – fermenting and pickling foods is an ancient practice and a great way to avoid waste.
3. **Save leftovers** – and actually eat them.
4. **Eat the skin** – potatoes, carrots, apples and even chicken!
5. **Eat the yolk** – ditch the faddy diets and eat the whole egg!

FACT:
NEARLY A THIRD OF ALL FOOD PRODUCED IN THE WORLD ENDS UP IN THE BIN!

RAINFORESTS

...ARE DISAPPEARING AT AN ALARMING RATE. BUT HOW CAN WE HELP?

You don't need to trek through the Brazilian rainforest and chain yourself to a tree to do your bit. In fact, these three easy steps can be done from the comfort of your own home...

1. **Avoid palm oil!** This common ingredient in processed foods and other products is the BIGGEST cause of deforestation.
2. **Create a welcoming environment for migrating birds and insects.** A bird house or a bug hotel will be a welcome sight after a lengthy migration from South America.
3. **Support indigenous communities** – buy fair trade foods, crafts and goods where you can.

SMALL CHANGES CAN MAKE A BIG DIFFERENCE!

BUT IT'S NOT ALL BAD NEWS.
REED INVASION MAY BE HELPING THE PLANET...

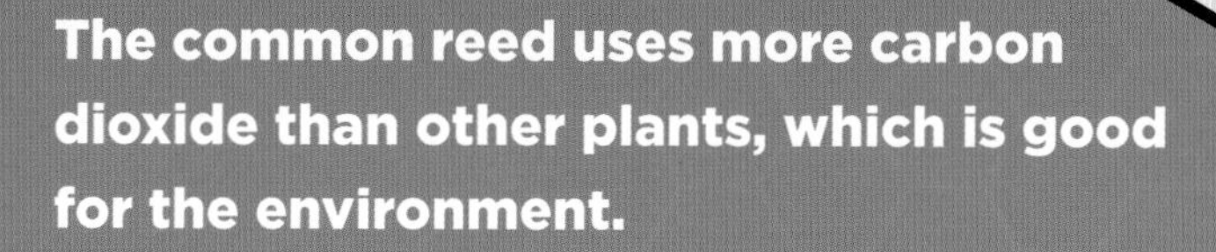

The common reed uses more carbon dioxide than other plants, which is good for the environment.

They also suck up pollutants from the soil, taking them out of our eco system.

GRASS INVASION!

VOLCANOS ARE GUILTY TOO!

IT'S NOT JUST HUMANS F**KING-UP THE PLANET!

BUT...

...before you start blaming volcanos for climate change, you should know the facts.

On average, volcanos emit 0.3 billion tonnes of carbon dioxide per year. Humans produce at least 100 times that.

And volcanos can't help spewing crap into the environment, **WE CAN!**

ERUPTION DISRUPTION!

GO VEGAN

EVEN IF IT'S JUST FOR VEGANUARY, OR ONE DAY A WEEK.

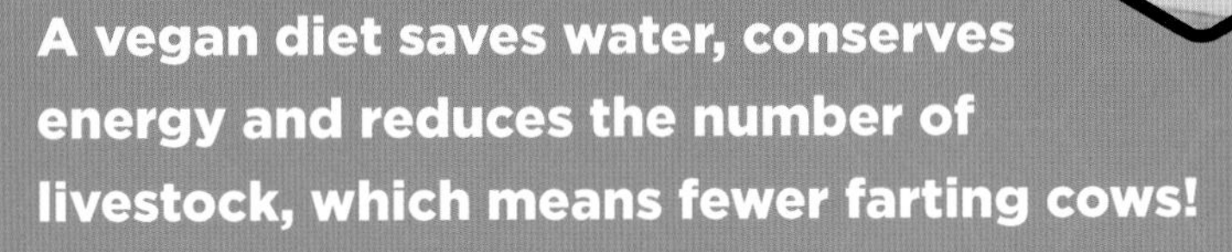

A vegan diet saves water, conserves energy and reduces the number of livestock, which means fewer farting cows!

It takes more than 2,400 gallons of water to produce 1 pound of meat, compared to growing 1 pound of wheat which only requires 25 gallons of water!

CALIFORNIA DREAMING

THE GOLDEN STATE IS KICKING CLIMATE CHANGE'S BACKSIDE.

California – the world's sixth largest economy – has published measures it's taking to cut carbon emissions by 40% by 2030.

CALIFORNIA LOVE

BAN THE BAG!

PLASTIC BAGS COULD SOON BE A THING OF THE PAST...

THREE American states – Hawaii, California and New York – have officially BANNED single-use plastic bags from circulation.

The rest of the world will hopefully soon follow....

FACT:

CALIFORNIA WAS THE FIRST STATE TO BAN PLASTIC BAGS BACK IN 2014.

CLIMATE CHANGE DENIERS DO EXIST. SOME EVEN HAVE ORANGE FACES AND LEAD THE FREE WORLD...

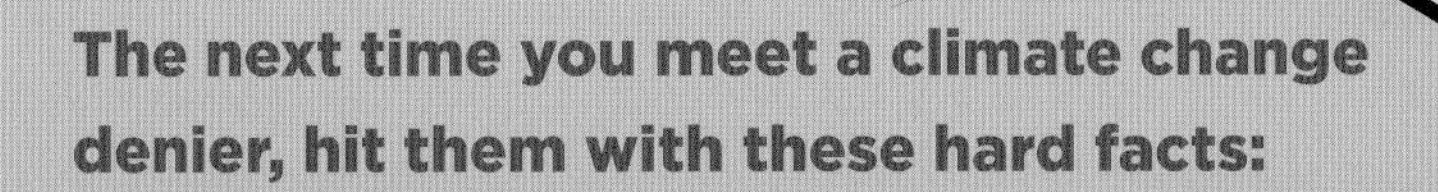

The next time you meet a climate change denier, hit them with these hard facts:

- Spring and summer are starting earlier each year, affecting plants and animals
- **The average sea level is rising by 3 mm per year**
- Arctic sea ice is shrinking
- **In England, average temperatures are between 0.5°C and 1°C higher than they were in the 1970s**
- Extreme weather events are on the rise...

FOOD WASTE AND EXPIRATION DATES

USE A LITTLE COMMON SENSE, PLEASE...

If your bread is covered with fur, growing legs and about to run away then obviously you shouldn't eat it.

HOWEVER, most food can still be consumed days if not weeks past its 'use by' date.

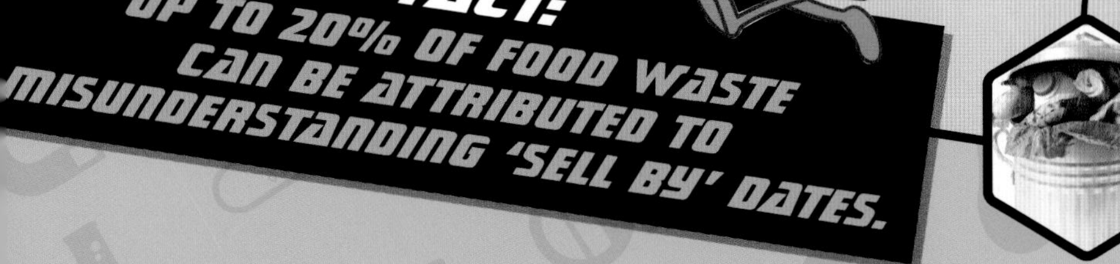

FACT:
UP TO 20% OF FOOD WASTE CAN BE ATTRIBUTED TO MISUNDERSTANDING 'SELL BY' DATES.

EARTH'S
NEW EPOCH
HUMAN BEINGS HAVE CHANGED THE WORLD
SO IRREVOCABLY; WE NOW LIVE
IN A NEW ERA OF HISTORY...

Our era in history is defined by the existence of atomic weapons, industrial gases, plastic pollution and carbon emissions.

A sobering fact, and even more reason to **GET INVOLVED** with the environmental movement.

THE AGE OF MAN – LET'S MAKE IT A GOOD ONE.

Photo Credits

Aleks47 / shutterstock.com
Mint Fox / shutterstock.com
Kotoffei / shutterstock.com
Afishka / shutterstock.com
Teguh Jati Prasetyo / shutterstock.com
Julia Tim / shutterstock.com
Darth_Vector / shutterstock.com
Memphisslim / shutterstock.com
Farhad Bek / shutterstock.com
Luciano Cosmo / shutterstock.com
Colorcocktail / shutterstock.com
Luciano Cosmo / shutterstock.com
Nexusby / shutterstock.com
Silvae / shutterstock.com
Officeku / shutterstock.com
CloudyStock / shutterstock.com
Paintings / shutterstock.com
Majeczka / shutterstock.com
ESB Professional / shutterstock.com
Artmim / shutterstock.com
Natasha Breen / shutterstock.com
AlenD / shutterstock.com
Happymay / shutterstock.com
Gyuszko-Photo / shutterstock.com
Vytas999 / shutterstock.com
Wayne0216 / shutterstock.com
Sunny Studio / shutterstock.com
Studiovin / shutterstock.com
JOJOSTUDIO / shutterstock.com
Graphego / shutterstock.com
Orange Vectors / shutterstock.com
AlenKadr / shutterstock.com
Alberto Masnovo / shutterstock.com
Alba_alioth / shutterstock.com
HQuality / shutterstock.com
Wake / shutterstock.com
Olha.vypovska / shutterstock.com
Kanittha Boon / shutterstock.com
KYTan / shutterstock.com
Anna Lshanksy / shutterstock.com
Coatesy / shutterstock.com
Coatesy / shutterstock.com
Billion Photos / shutterstock.com
Squarelogo / shutterstock.com
Rich Carey / shutterstock.com
Daisy Daisy / shutterstock.com
Melanie Hobson / shutterstock.com
JakubD / shutterstock.com
MicroOne / shutterstock.com
FXQuadro / shutterstock.com
Studiovin / shutterstock.com
Chaiyapruek youprasert / shutterstock.com
ViDi Studio / shutterstock.com
Maloff / shutterstock.com
Yuriy Golub / shutterstock.com
VECTOR FUN / shutterstock.com
Hedgehog94 / shutterstock.com
May-Helen Mogstad / shutterstock.com
Holli / shutterstock.com
Monica Goebel / shutterstock.com
David Pereiras / shutterstock.com
Milos Moncilovic / shutterstock.com
Aaron Amat / shutterstock.com
Just Dance / shutterstock.com
Firn / shutterstock.com
Anton_Medvedev / shutterstock.com
Greyboots40 / shutterstock.com
Frantic00 / shutterstock.com
Timon Goertz / shutterstock.com
FXQuadro / shutterstock.com
AlenKadr / shutterstock.com
Marcus Gier / shutterstock.com
nito / shutterstock.com
Javid Kheyrabadi / shutterstock.com
Dagmara_K / shutterstock.com
Igor_Koptilin / shutterstock.com
oleschwander / shutterstock.com
Rich Carey / shutterstock.com
Mix and Match Studio / shutterstock.com
Ian Murdoch / shutterstock.com
2M media / shutterstock.com
Paapaya / shutterstock.com